The Song of the Stars

THE SONG OF THE STARS

**Starry Skies,
Anishinaabe Stories,
Scientific Insights,
and More!**

ROBERT ANIMIKII HORTON

ILLUSTRATIONS BY KC OSTER

AEVO UNIVERSITY OF TORONTO PRESS

Aevo UTP
An imprint of University of Toronto Press
Toronto Buffalo London
utppublishing.com

ISBN 978-1-4875-6415-5 (cloth)
ISBN 978-1-4875-6417-9 (EPUB)
ISBN 978-1-4875-6416-2 (PDF)

Library and Archives Canada Cataloguing in Publication

Title: The song of the stars : starry skies, Anishinaabe stories, scientific insights, and more! / Robert Animikii Horton.
Names: Horton, Robert Animikii, author
Description: Includes bibliographical references.
Identifiers: Canadiana (print) 20250249294 | Canadiana (ebook) 20250249340 | ISBN 9781487564155 (cloth) | ISBN 9781487564162 (PDF) | ISBN 9781487564179 (EPUB)
Subjects: CSH: Anishinaabeg. | LCSH: Stars. | CSH: First Nations cosmology. | CSH: First Nations astronomy.
Classification: LCC E99.C6 H67 2025 | DDC 520.89/97333 – dc23

Printed in Canada
Cover design: Will Brown

We wish to acknowledge the land on which the University of Toronto Press operates. This land is the traditional territory of the Wendat, the Anishnaabeg, the Haudenosaunee, the Métis, and the Mississaugas of the Credit First Nation.

University of Toronto Press acknowledges the financial support of the Government of Canada, the Canada Council for the Arts, and the Ontario Arts Council, an agency of the Government of Ontario, for its publishing activities.

Canada Council for the Arts | Conseil des Arts du Canada

ONTARIO ARTS COUNCIL
CONSEIL DES ARTS DE L'ONTARIO
an Ontario government agency
un organisme du gouvernement de l'Ontario

Funded by the Government of Canada | Financé par le gouvernement du Canada | Canada

FSC www.fsc.org MIX Paper | Supporting responsible forestry FSC® C103567

For Neil and Mandy.

Contents

Preface

This book initially began as a conversation between the good people from Kay-Nah-Chi-Wah-Nung Historical Centre and me in 2019. Following long talks sharing Anishinaabemowin's connection to concepts in our universe (and inclusion of such into speaking engagements at the centre's roundhouse), staff thought the creation of a resource to share such insights would be a positive contribution to our region. The idea of looking at events in our universe from different perspectives (blending ancient knowledge held by family and teachers, deep-dives into our language shared by language mentors, and many years of personal study and fascination with science) was something they immensely welcomed. With their encouragement and supportive words, this book was authored and completed in a winter cabin on the shores of Longbow Lake – not far from Wazhashkonigaming.

Also called the Manitou Mounds, the Kay-Nah-Chi-Wah-Nung Historical Centre is located near Stratton, Ontario. Designated a National Historic Site in 1969, it is the site of the highest concentration of traditional burial mounds in Canada. The land where the centre is located, Ginwaajiwanaang (also known as Long Sault Rapids), is where my family long made their home prior to the relocation to Manidoo Baawitigong (Manitou Rapids) in 1914. Knowing the book's concept began in the same place where family cared for the same teachings and language during their lives makes me grateful to be part of this project during ours.

Introduction
In Awe of the Awe-Inspiring

Since the earliest days of human memory, countless generations have turned their eyes to the skies in wonder and awe – and we still do.

As time progressed, watchful eyes from across the globe began to connect patterns above, understand the stars' connection to cycles of events, and (in turn) connect stories and teachings to coming generations with anticipating hearts.

This book is a very special vantage point intended to connect us to our ancestors who watched the very same wonders of our universe that we still see today. By looking to the very same skies (and perhaps with a better understanding of their stories and perspectives), we can stand beside them with the same wonder, and time can melt away.

In these pages, you will discover a multitude of star constellations, stellar bodies, celestial events, and interwoven teachings (and stories) of the Anishinaabe. Also, several teachings about celestial events, stars, and constellations from other cultural perspectives across the globe will be briefly acknowledged. Finally, scientific insights will be shared to supplement such teachings and at times provide new understandings to these awe-inspiring events in our solar system and beyond.

Why Cultural Stories and Teachings?

Like countless cultures across the globe, the Anishinaabeg have always looked to the skies and observed a seemingly endless ocean of stars. Understandings of what was observed in the skies shone through one's living culture like a prism while shaping long-standing connections to these sources of wonder. These perspectives have been captured and expressed within the Anishinaabe language. Intimately connected to the cycles of life and the land, untold generations used observation driven by passionate inquiry to better understand the environment and life processes (such as when to harvest and prepare for the winter season). The same committed observations led to more questions, which led to further learning and understanding. Over time, stories and teachings would be woven into what was witnessed above as philosophical, metaphysical, moral, and spiritual insights were held firmly at the core of each. Informed by wisdom of countless lived experiences, collective memory, and spiritual ceremonies, these teachings (blended with stellar patterns) would successfully wrestle with many of the major questions, insights, and perspectives shared by the human family across the world. Such spiritual and philosophical insights often reach beyond the physical and material realm into metaphysics and ethics shared through story, ceremony, and wisdom – and still do. Stories and teachings acknowledge spiritual beings and helpers who share our Earth. The language of Anishinaabemowin and teachings connected to the skies above ensure that the ethics and philosophies attached to such narratives and stories survive. Sophisticated and purposeful, these star patterns and celestial events (and the teachings connected to them) ensure that responsibilities and values carried by Anishinaabeg (as well as time-tested ways to navigate challenges also shared by the wider human family) continue far into the future…into today and many tomorrows.

Why Scientific Insights?

Only a few centuries ago, a very successful method was devised and developed across the ocean with the goal of increasing accurate understandings (and limiting inaccurate misunderstandings) of the natural, material, physical world. Based upon study and evidence (as well as tools that include testing and peer review), the scientific method has brought matters beyond observation and inquiry into a devised method to meet such goals. It is defined by rigor and analysis, and has helped our human family improve everyday life, discover new solutions to new challenges, and unlock many mysteries of the cosmos by means of testable (and predictable) explanations. Many cultures have contributed to modern science's humble beginnings, and science now belongs to (and is utilized by) the human family to explain objective, physical reality and increase well-being. For something to be considered modern science, the scientific method must be followed. This includes the process of observation/questioning, research, hypothesis, testing and experimentation, analysis, and reporting conclusions, followed by peer review (which often leads to re-testing). This method has been utilized across the globe and has proven useful in further discoveries in our world and beyond. Over time, specific fields of science have developed. Following consistent use of the scientific method (while confirming earlier findings), clarity on matters in the natural world improves and becomes clearer. Over time (and in a symbolic comparison), it is almost as if we're steadily increasing the visual resolution of what we attempt to see better and understand. Relevant to this book, the field of astronomy has been massively insightful.

It's important to understand that the term "theory" in the scientific realm is often quite misunderstood. A scientific theory does not mean an aimless guess, but rather an explanation of the natural world that has been tested and corroborated through the scientific method, utilizing observation and methods of measurement, and repeated results.

Why Include Both Cultural Teachings and Scientific Insights?

Both cultural teachings and scientific insights can be intensely valuable and can be incredible (and memorable) sources of absolute wonder, wisdom, reflection, and insight.

Stories and cultural teachings not only connect us to generations that came before, but also provide those living today with rich perspectives, contributions, and connections made between our place in the universe and embedded philosophies, ethics, and values. By means of language and teachings, these have been woven into the wonders of life across generations. These stories, narratives, and teachings dance with metaphysics in order to provide insights about ways to pursue good lives.

Another source of true wonder, science provides rigorous, systematic, and testable explanations to our objective, physical, and material reality. Because of its constant growth, further discovery and solutions to new challenges can be navigated.

On a fundamental level, perhaps two very different questions are being answered by each respective approach. It could very well be that cultural teachings inform us about good ways to walk, relate, and pursue deep metaphysical truths and meaning while in this living and mysterious universe, while science informs us of the nature of the universe we walk through during our lives.

It should be noted that knowledge, facts, evidence, and objective reality should be approached with equal parts confidence and humility. Although science has given us an incredibly robust understanding of our physical reality, there is always still more to know. The pursuit of knowledge is forever. Likewise, it should be noted that belief, opinion, and subjective perspective should also be approached with equal parts wonder and humility. In this realm, it is never "the way," but "a way." However, with an objective understanding of our world, several beliefs and perspectives may be confirmed and championed. Likewise, philosophical understandings can bring awe and personal rapture to the wonders of the universe.

It is absolutely crucial to acknowledge the knowledge and revelatory insights gained by means of our ceremonies, as well as our spirit helpers and the true-blue gifted carriers of such ceremonies. This added dimension of reality and experience (as well as our helpers that guide us and provide insight) should never be overlooked. Such a dimension and source of inspiration and insight exists across the globe, across cultures, across languages, and is appropriately given reverence in these pages. Also, efforts to explore beyond surface translations into deeper esoteric meanings of Anishinaabemowin (at the heart of concepts shared) are shared with great care. Such esoteric meanings are alive in the heart of our language and have been cared for, protected, and expressed across generations.

This resource is meant to highlight the common ground that cultural beliefs and scientific knowledge share, where they depart, where one may inform the other, and to reveal the sacred chords they each play in the eternal song of the stars.

Respect and Reciprocity

Great care and mindful respect have gone into creating this book. While utilizing generations of stories and teachings from our families along the Rainy River, a conscious choice to limit the depth of traditional teachings was carefully undertaken. As several lodges (and very respected spiritual leaders) are truly the prime source of in-depth and complex cultural teachings, I have chosen to share things only to a certain (and perhaps limited) depth and degree. For instance, the long-standing connections between the stars above, rock paintings throughout our lakes and waterways, and the language that defines them will not be explored in an in-depth manner. "Going to the source" is key for those who wish to take their learning further.

With sincerity and reciprocity, this book is meant to be a meaningful contribution to our communities.

The Future

Considering recent events such as the development and launch of the James Webb Space Telescope, robotic explorers sent to Mars, and the activation of the European Organization for Nuclear Research (also known as CERN), the future of our knowledge systems about the universe seems very bright indeed.

Acknowledging the revitalization of Anishinaabe language, the continuation of connected teachings, philosophies, ethics, and perspectives to be passed successfully to coming generations seems to be optimistic, as well.

We, collectively, are continuing to learn and grow – together.

It's clear that our vantage point on Earth is one of awe, and I hope it is further brought to life in these pages.

Whether cultural understandings from a vast diversity of cultures and societies across the Earth or recent, objective understandings by the human family about our world and universe, each produces a unique tone in the grand scheme of the story of understanding.

Each is a tone, a note, in the song of the stars.

Anishinaabemowin Sounds

As Anishinaabemowin will be used in many areas of this book, the following sounds will assist with pronunciation.

Long Vowel Sounds	Examples
aa – "aah" like in f**a**ther.	jiim**aa**n (boat)
e – "ay" like in caf**é**.	ikw**e** (woman)
ii – "ee" like in fr**ee**.	j**ii**maan (boat)
oo – "oo" like sh**oe**.*	b**oo**zh**oo** (hello)

* At times (in some regions), oo may sound like "oh" in boat

Short Vowel Sounds	Examples
a – "uh" like in sh**u**t.	m**a**nidoo (spirit)
i – "ih" like in t**i**p.	n**i**b**i** (water)
o – a short "o" sound like in c**oo**k.	**o**pin (potato)

New Consonant Sounds	Examples
ch – like in **ch**op.	mi**ch**aa. (It is big.)
sh – like in **sh**op.	zhii**sh**iigwan (shaker)
zh – like in trea**s**ure.	boo**zh**oo (hello)
' – like the short pause like uh-oh.*	dewe'igan (drum)

* This short pause (represented by an apostrophe) is called a glottal stop.

The English consonant sounds of c, f, l, q, r, u, v, and x are not used in Anishinaabemowin. However, the remaining consonant sounds used in Anishinaabemowin are consistent with their English equivalents.

Nasal Vowel Sounds

The following sounds are said through the nose. The "n" sound will not be pronounced as the tip of your tongue does not touch the back of your teeth.

Sound	Examples
aanh	banaj**aanh** (a nestling)
enh	nimis**enh** (my older sister)
iinh	awes**iinh** (a wild animal)
iiny	abinooj**iiny**ens (a baby)
oonh	giig**oonh** (a fish)

Nasalized Vowel Sounds

All vowel sounds pronounced before the following three sounds are said through the nose. The "n" sound is not pronounced as the tip of your tongue will not touch the back of your teeth.

Sound	Examples
nz	m**oonz** (a moose)
ns	ikwez**ens** (a girl)
nzh	oshk**anzh**iin (his/her fingernail)

Aki
The Earth

On the surface, the word Aki refers to the Earth.

However, an ancient understanding of the word Aki is "everything with life energy," invoking the Earth and everything connected to this very special place. This acknowledges that the Earth is a massive source of life which supports the entire web of species in relation to one another.

Across recent generations, Aki has also been used to describe land, the ground, or even a country.

Many other words in the Anishinaabe language that contain the sacred word portion of "kamig" refer to a concept, location, event, or idea with spiritual bearing in relation to the Earth.

Beautiful cultural stories and teachings from across the globe speak of a conscious process of a creation of the universe, the planets, sacred directions, the seasons, the Earth, as well as the web of life that calls the Earth home.

To the Anishinaabeg, the Earth is identified as a collective Mother and is part of a larger celestial family that includes the Grandmother Moon, Grandfather Sun, and others.

The Earth is acknowledged as the most crucial form (and source) of life in the web of existence. No form of life, from vegetation, animals, or humans, can exist without her. In fact, it is said that humans are last in the chain of importance among the relationships.

Although rivers and lakes have served as a means of transportation for the Anishinaabeg (and countless others) throughout history, these waterways (as well as the groundwater) are said to be Mother Earth's lifeblood.

In this intricate web of life, the way we impact the Earth affects all forms of life – including ourselves.

According to the Anishinaabe teachings, the Creator (using four sacred elements of the Earth) blew into a sacred Miigis shell and created the original beings who were lowered down upon the Earth in different regions.

It is said that there are very sacred layers above and below the Earth which hold both physical and spiritual significance.

Niiyo-biitookamigishimind refers to the Earth and the structural and spiritual layers within.

The rocks (asiniig) are said to be our oldest relatives here on Earth and have experienced more than any form of life present. This is why they are often referred to as Grandfathers.

Did You Know?

There are rich and robust cultural teachings across the globe that tell stories about the Earth.

These include the creation story of Heliopolis in Ancient Egypt, the story of Tepeu and Gucumatz among the Maya, the world emerging from a sacred shell in Polynesia, and Pangu to the Chinese Daoist.

When viewed from space, the abundance of water makes Earth very distinct in our solar system. The oceans contain 75 percent of Earth's water. Another 2 percent can be found within glaciers and ice sheets.

Less than 1 percent is present in groundwater and held within rocks. The rest is found in the atmosphere, lakes, streams, creeks, and rivers. Water on the surface of Earth, underground, and in the air is collectively called the Earth's hydrosphere.

Earth is full of life. There are as many microbes in one teaspoon of the Earth's soil as there are humans currently living on the continent of Africa today.

The biodiversity on Earth is massive. Roughly 1.5 million species have been included and described in scientific literature. Researchers from the University of Arizona have estimated there are approximately two billion living species on Earth.

Ninety-nine percent of all species that have ever lived on Earth are now extinct.

Human beings are genetically related to life here on Earth. A human has 99.9 percent genetic similarity to another human in our DNA, as well as 96 percent genetic similarity to a chimpanzee, 90 percent to a cat, 85 percent to a mouse, 80 percent to a cow, 61 percent to a fruit fly, 60 percent to a chicken, and 60 percent to a banana.

Nearly 65 million years separates the extinction of the dinosaurs and the appearance of people.

The Earth is not a perfect sphere. It is known as an oblate spheroid – similar to a somewhat squashed sphere. The swell at the equator is due to the Earth's spin, the effect of gravity, and centrifugal force.

At the equator, the circumference is over 40,000 kilometers.

Evidence shows that the age of the Earth is 4.54 billion years old.

Above the Earth's surface, there are four main layers to the atmosphere which are distinguished by features such as temperature. Below the surface, there are distinct layers to the structure of the planet including the crust, the mantle, the outer core, and the inner core. All contribute to ensuring life is maintained and protected on the planet.

The Earth's distance from the sun, physical properties, and history related to geology have allowed life to appear and to thrive.

Giizis
The Sun

Just as the word Giizis conveys a being that radiates life energy, the sun is the most important source of energy for life on the Earth.

Anishinaabe stories and teachings acknowledge the sun as a Grandfather.

When spoken about, the name Gimishoomisinaan Giizis ("Our Grandfather Sun") reveals something special embedded in language. This form of Grandfather is inclusive of the human family and all life on Earth. The sun shines on all equally.

In some regions, this concept is conveyed as Gigichimishoomisinaan Giizis.

Biidaaban.	"The dawn is coming."
Waaban.	"It is dawn/daybreak."
Mooka'am.	"It is sunrise." (Literally, "he/she is rising.")
Bangishimo.	"The sun is setting." (Literally, "he/she is sitting down.")

There are many Anishinaabe teachings that are centered upon (or that include) the Grandfather Sun.

When Migizi (the Bald Eagle) prevented the Sun being blown out, Giizis was identified as a being that plays a massive role in providing life energy and survival to all beings.

Also, it is said that Grandfather Sun and Grandmother Moon carry their own responsibilities, but there are times they work together (such as during an eclipse).

One notable event involving sunlight is known as parhelion. Also known as a sundog, this atmospheric optical phenomenon appears as patches of light to the right, left, or both sides of the sun. This occurs when sunlight refracts through ice crystals in the atmosphere. Waabizhe'o is a very ancient Anishinaabe word for the phenomenon of sundog (conveying the relationship occurring between intense light, water in a particular form, and the feel within the observer). The word for a pine marten is waabizheshi, acknowledging a luminous sundog upon one's chest (appearing as lighter colored fur) – *the being carrying the sundog.*

Did You Know?

Throughout history and across the globe, countless cultures have connected stories and teachings to our Sun.

These include Helios of Ancient Egypt and the chariot driven across the sky, as well as solar deities such as Lugh to the Celts, Tonatiuh and Huizilopochtli to the Aztecs, and Shamash to the Sumerians.

This makes sense because of the Sun's consistent and prominent role in life on Earth.

The sun is our closest star and is at the center of our solar system. It is just one of the many countless stars we see in the night sky.

A sphere of intensely hot plasma, it contains 750 times the mass of all the planets in the solar system.

The sun is currently 4.6 billion years old.

Energy is generated at the center of the sun by nuclear fusion (as hydrogen nuclei convert to helium). Every second, this process fuses over 500 million metric tons of hydrogen.

The temperature at the core of the sun can reach 15 million degrees Celsius. However, the surface is over 5,500 degrees Celsius.

Light from the sun takes eight minutes to reach our eyes on Earth. This means when we see the sun, it is not where it appears to be. Instead, it is eight minutes ahead of where it appears to be. We merely see where it was eight minutes ago.

The light contains all colors together, which is why it appears as white to our eyes during the day. What about during sunrise and sunset? There is a reason why the sun appears yellow, orange, or red at these times. It is lower on the horizon and there is much more air and atmosphere to pass through. These molecules scatter the violet and blue light away from our eyes.

The sun's explosive nuclear energy pushing outward (as well as the effect of its own gravity pulling it inward) is an amazing example of balance demonstrated in our universe. This will continue until toward the end of the sun's life when its hydrogen becomes depleted.

Gichi-giizis
The Solar Eclipse

A solar eclipse is one of the most awe-inspiring moments in the natural world.

Depending on the region, the Anishinaabe name of the celestial event when the moon moves between the Earth and the Sun may be different.

To the South, makadewaabikizi ("he/she is in eclipse") is used to describe this moment.

To the East, some may describe it as containing the word core of Gibiske, pertaining to an obstruction from sight.

Along the Rainy River, Gichi-giizis is an ancient term acknowledging a solar eclipse by describing the intense light and life energy when the sun slowly reveals itself from behind the moon.

As the name Giizis describes the radiating of life energy, one can appreciate how much energy reaches us by witnessing the dazzling corona of the sun that is clearly revealed in the moments of a total solar eclipse. As the sky is darkened, one can also truly understand the sun's true brightness in the moments before, during, and after the moon obscures the sun from our sight.

There are many teachings and stories connected to this celestial event. As the sun is said to be a Grandfather (and the moon is a Grandmother), it helps to demonstrate the balance between the male and the female in the cycles of nature and who are responsible for life. A solar eclipse is a time when they both come together to honor their relationship and responsibilities to each other. Just as the Grandfather and the Grandmother have their own tasks in the sacred order of the natural world, it is a reminder that men and women also have responsibilities to one another to honor and fulfill – but there is always equality and balance.

This insight and teaching was given to Nanaboozhoo as he made his way across the plains in search of his father long ago.

Did You Know?

Anishinaabe teachings are not the only place stories connected to the solar eclipse have been found in our collective history of the human family.

The Chinese once believed that an invisible dragon had a role in this celestial event, just as the Greeks told stories of deities who were dissatisfied and foretold of destruction to come.

Indigenous Peoples to the North spoke of a quarrel between Malina and Anningan – prominent spirits connected to the Sun and the Moon.

The distance between the Sun and the Earth is roughly 400 times the Moon's distance from the Sun. At the same time, the Sun's diameter is roughly 400 times larger than the Moon's diameter. This is why the Sun and the Moon appear to be similar in size from the Earth and why a solar eclipse can happen with such beauty.

During a solar eclipse, it gets dark on Earth because the moon casts a shadow on a particular region.

The shadow cast during a solar eclipse has three parts. They are known as the antumbra, penumbra, and umbra. Each part of the shadow has a different degree of darkness.

There are four types of solar eclipses. They are called total, annular, partial, and hybrid. Each has different, distinctive features and differs by how fully the sun is obscured by the moon.

The narrow route of shadow where you can see a total solar eclipse is called the path of totality.

It is important to take safety measures if you plan to view a solar eclipse (such as a safe solar viewer). Looking directly at the sun can cause damage to the eyes.

Total solar eclipses occur somewhere on the Earth roughly every 18 months and (on average) between 360–410 years for a recurrence in a certain location.

Naawakwe
Solar Noon

Naawakwe ("it is noon") is a very special phrase that describes a particular time of day.

This doesn't necessarily mean twelve o'clock on the dot. It is another type of noon that is interwoven with the movement of the Earth in its relationship to the Sun.

Just as the word Naawakwe conveys a time when the peak of life-giving energy radiates the most intensely from the sun, it speaks of the time of day that Gimishoomisinaan Giizis (our Grandfather Sun) appears at the highest point in the sky above where one is located.

In any area, this occurs exactly halfway between sunrise and sunset.

For generations in the Rainy River region (as well as elsewhere), it is a significant time of day when offerings are put out and feasting spiritual helpers are done. According to regional teachings, this is done because there is an intense peak of life energy cast toward the Earth. By means of ceremonial protocol and made possible by use of our language, we can work with this energy. Anishinaabemowin informs us of this:

Naa pertains to a sound to call attention to someone or something.
Wa (a derivative of Waa) pertains to visible light (containing intense life energy being radiated from the sun).
Kwe pertains to a particular moment of time.

Did You Know?

Although civil time (using a global system of time zones made of one-hour blocks) greatly helps us organize our world and keep our schedules in sync, it isn't perfectly fine-tuned to the rotation of the Earth in relation to the Sun. Just as noon (using civil time) separates morning and afternoon in a particular time zone, solar noon (using solar time) occurs when the sun crosses the local meridian and appears at the highest point in the sky.

This specific time can be different from the noon we see on our clocks.

At times, astronomers refer to solar noon as "local apparent noon." As the Earth rotates, solar noon will be different from one point on the globe to another. Depending where you stand on the Earth, solar noon happens earlier in locations to the east of you and later in locations to the west of you.

At a location's solar noon, the shortest shadows are cast.

Ma'iingan Omiikana
The Sun's Ecliptic

When observing from the Earth, there is an imaginary line that the sun takes across the sky during each day over the course of the year. This can be described as the path of the sun.

The name of this path in Anishinaabe teachings is Ma'iingan Omiikana (or in some regions, simply Ma'iingan Miikana) – the Wolf's Trail.

Ma'iingan (the wolf) has a very important role in traditional teachings and stories. When Original Man (who was named Anishinaabe by the Grandmother Moon) was created, one of his first tasks was to travel to Earth and name all that he saw before him. It is said that this is what gave rise to Anishinaabemowin (the Anishinaabe language), as well as Wiindaawasowin (the Naming Ceremony). As Anishinaabe traveled the land, he noticed that life came in pairs. Upon asking for a companion of his own, Ma'iingan was sent to walk the land and explore all the beautiful places with Anishinaabe. After time passed and this task was completed, Anishinaabe and Ma'iingan were told to separate their paths from that day forward. This set the foundation for protocols regarding dogs (descendants of wolves) to keep a distance from the vast majority of ceremonies today.

Even though the paths of Anishinaabe and Ma'iingan separated, the destiny of one is connected to the other. It is said that what would occur to one would occur to the other in the future.

Many believe this explains why between human beings and dogs (who are descended from wolves) maintain such a positive relationship today.

The Ma'iingan Omiikana is the path that Ma'iingan has walked (and continues to walk) through the journey of space and time.

Did You Know?

How this path appears from Earth is majorly dependent on the time of year.

It is the plane of the orbit of the Earth around the Sun.

When one pays attention to the sky, the location of the ecliptic shifts a bit seasonally.

Because the Earth's axis rests at 23.5 degrees, it causes the sun to appear higher in the summer and lower in the winter.

The ecliptic serves as a very important reference point for astronomers and stargazers.

There is a prominent place of the Sun's path through the sky and the stories of the constellations within the Ancient Zodiac.

Aadwaa'amoog
Orion's Belt

When you look up at the night sky, one of the most recognizable constellations is Gaa-bibooniked, the One who Works with the Winter. This constellation is also known as Biboonikewinini, the Wintermaker, commonly known as Orion.

Easily identifiable to the naked eye, three bright stars rest near the center of the constellation.

Another ancient Anishinaabe story speaks of these three stars as three individual beings or spirits making their way across the night sky. Aadwaa'amoog expresses that they are traveling together.

In the context of an ancient sky, it implies that a boat or canoe is the means of travel that the three beings are using to get to their destination. Aadwaa'amoog's pattern constellated in the sky (three beings travelling by canoe) has manifested via rock paintings in the beautiful Rainy Lake region. When inverted, this conveys the three beings are travelling "on the other side" rather than in the world that living beings directly experience each day.

In some regions, these three stars are described as three parts of the construction of the canoe they are using to paddle across the sky.

This method of travel is very significant throughout Anishinaabe history. This is a wonderful example of how cultural experiences (such as the construction and use of a birchbark canoe) can shape ancient perceptions of constellations in the night sky (and can vary according to region and by time).

Did You Know?

The window between November and February is the prime time to see this constellation in the southwestern sky.

During the beginning of the second week of January, a notable time to view it very clearly is roughly four and a half hours after sunset.

The Arabic names Alnilam, Mintaka, and Alnitak are the common names of these prominent stars in the night sky that make up Orion's Belt.

Because of their positions (which appear from our vantage point as an evenly spaced straight line), it is easy to spot.

Throughout history, many cultures and civilizations have passed stories down through generations about this trio of stars. The Greeks told stories of a great hunter with the three stars as his belt. In Ancient Egypt, the three stars were connected to Osiris. In New Zealand, the Māori refer to it as Tautoru, meaning "string of three." In different areas of the world, such as the Philippines and Mexico, some describe them as the Three Marys, whereas others tell stories of Three Kings.

Because of its widely recognized pattern of stars in the night sky, it is known as an asterism in astronomy.

Alnitak (the eastern-most star of the trio) is a triple star system and is roughly 1,260 light years from Earth.

Alnilam (the middle star of the trio) is roughly 2,000 light years away from our home, the Earth. It is known as a blue supergiant.

Do you think our local star, the Sun, is bright? Alnilam is hundreds of thousands of times more luminous. Mintaka (the western-most star of the trio) is a multiple star system at roughly 1,200 light years away from the Earth.

Waawaate
The Northern Lights
(Aurora Borealis)

Have you ever looked up at night and witnessed different colors illuminating the sky and racing from one side of the sky to the other? These are the Northern Lights.

Waawaate is more than a name. Rather, it is describing an action taking place in the world around us – "it is the Northern Lights." It is also a description of the beauty and the glow from above. "Waa," originating from Waas, describes a striking visible shine or glow. When it is repeated, such as "Waawaa," we note both the brilliance of the illumination and the oscillation of the light fading in and out, continually becoming visible and then not. "Aate" notes that this wonder has light as its originating source. This provides a linguistic picture of brilliant, undulating luminescence which fades in and out, originating from a brilliant source of radiance.

To the South and West, Jiibayag Niimi'idiwag is the description given to this celestial event. This means that the spirits of people (those who lived before) are dancing with one another.

Although the name of the event is different in the Rainy River region, it shares consistent stories and teachings – spirits of people dancing brilliantly across the night sky.

Did You Know?

The Aurora Borealis event has fascinated people all over the world for thousands of years. Its name has roots in Greek and Roman mythology. Aurora is the Latin word for dawn and was the ancient Roman name of Eos, a deity who was the personification of the dawn. Boreas was a Greek deity of the North Wind. The Greeks told stories of Aurora racing across the sky during the early morning. In her colorful chariot, she shared the news of a new day with her family.

The Italian astronomer Galileo Galilei coined the term for this event.

In ancient Scandinavia, stories were shared that the Norse deity Ullr illuminated the skies on the longest nights of the year. Some in Norway told stories of the light being produced by the armor and shields of the warriors chosen by the Valkyrie. It was reported as a red glow in the night sky on a Babylonian tablet when King Nebuchadnezzar II ruled. Pliny the Elder of Rome described the event in the first century as a yawning of the night sky.

This event truly has a light as its origin source. Solar wind steadily streams off the Sun in all directions from its brilliantly hot corona, and solar particles travel through space at millions of miles per hour. More solar particles from the Sun can also be thrown during massive eruptions known as coronal mass ejections.

By the time they reach the Earth's atmosphere, these particles have traveled over 93 million miles. A majority continue past Earth. However, those that enter the Earth's upper-atmosphere collide with atoms and molecules which give off light in an excited state. The Earth's magnetic field guides them toward the North and South poles.

Those that find themselves pulled toward the South pole are called the Aurora Australis – the Southern Lights.

Jiibay Omiikana
The Milky Way

If you've ever been looking upward at night and seen a hazy, wispy band of light that crosses the sky, you've seen the Milky Way. Stretching from one side of the sky to the other, it is full of so many stars that it is nearly impossible to tell them apart. The further one gets from light sources in towns or cities, the more clearly one can see it on a clear night.

This is Jiibay Omiikana (The Spirit's Road), also called Jiibay Miikana, Jiibaykana in some regions, and Jiibay Ziibi (The Spirit River) to the east.

The language used is not only very descriptive, but also very telling of who this road is for in traditional stories and teachings. As opposed to Manidoo (a supernatural being), Jiibay describes the spirit that has resided within a being's physical vessel – the spiritual identity of a being that has walked this Earth, disembodied.

Since ancient times, stories among the Anishinaabeg identify this band of light stretching across the entire night sky as the road (or in some regions to the east, a river) by which human spirits enter and exit this world where we live our lives to learn, experience, and fulfill responsibilities. Many teachings and stories describe this road that human beings travel, before or after their lives on earth.

Did You Know?

There is much more to the Milky Way than what we can see from our vantage point here on Earth. It is our galaxy, and because we are in it, it is difficult to grasp the immensity or see it fully. It is a massive, flat, disk-shaped galaxy with stars, planets, gas, dust, and more.

Evidence supports that the Milky Way is 13.6 billion years old. It is older than our planet and solar system, as well as nearly as old as the universe itself.

There are four main spiral arms to the Milky Way galaxy named Norma and Cygnus, Sagittarius, Scutum-Crux, and Perseus.

Our Sun is located in a spur (also called a minor arm) called the Orion Spur.

What we see as a hazy band of light stretching from horizon to horizon is a very, very, very small portion of the Milky Way galaxy. It would take over 100,000 years for light to go from one side of the Milky Way to the other. Recent studies suggest this number could actually be double because of newly observed stars among the galaxy's rim.

It has been said that at the center of the Milky Way galaxy is a black hole. A black hole is a very dense point in space where an astronomical body has undergone a gravitational collapse. The gravity is so strong in these places that not even light can escape. Astronomers can detect possible locations by watching the activities of stars in orbit around areas that cannot be seen.

This supermassive black hole of more than four million solar masses is called Sagittarius A*. It is the Galactic Center, the rotational point of the Milky Way galaxy.

Gookomisinaan Dibiki-giizis
The Moon

The moon has a great cultural significance.

When spoken about, the name Gookomisinaan Dibiki-giizis ("Our Grandmother Moon") reveals something special embedded in language. The Grandmother is inclusive to the human family and all life on Earth.

In some areas of our region, some use the terms Dibiki-giizis or Dibiko-giizis only.

In Anishinaabe teachings, it is revealed that the Grandmother who named the being known as Anishinaabe resides there and has a profound effect on the waters and looks after the women who call the Earth home.

For countless generations, feasting ceremonies have been conducted by women to honor Gookomisinaan during each lunar cycle when the moon is full. This special ceremony is still done today.

Mooshkine-giizis is a common name for a Full Moon.

Aabita-giizis is a common name for a Half Moon.

The Moon doesn't radiate her own light. Instead, she reflects the light created by the Sun. The reflection of light is often described as the Grandmother speaking on behalf of the Grandfather. Her name contains awareness of this, conveying life energy that is given a trajectory or angle for a particular time. Our Grandmother Moon also supports Anishinaabeg by reminding us of natural processes and distinctive seasonal changes to come.

The names of each period acknowledge what is occurring in the environment at that specific time. In fact, this practice of Indigenous Peoples giving names to temporal changes in the surrounding environment predates the Julian and Gregorian calendars.

Aabita-bibooni-giizis – The Halfway Through Winter Moon (January)
Migiziwi-giizis – The Eagle Moon (February)
Aandego-giizis – The Crow Moon (March)
Maango-giizis – The Loon Moon (April)
Zaagibagaawi-giizis – The Bud Bursting Moon (May)
Ode'iminikewi-giizis – The Strawberry Picking Moon (June)
Miinikewi-giizis – The Blueberry Picking Moon (July)
Manoominikewi-giizis – The Wild Ricing Moon (August)
Waatebagaa-giizis – The Bright Leaves Moon (September)
Binaakwiiwi-giizis – The Falling Leaves Moon (October)
Gashkadino-giizis – The Freezing-Over Moon (November)
Bibooni-giizis – The Winter Moon (December)

It is important to remember that these natural processes of the land, which correspond with the moon, are not confined to the month names of September, October, and others. The month names to the right (in parentheses above) are merely a way to organize it and to get a general idea when they occur.

For instance, Waatebagaa-giizis may begin at the end of August or end at the beginning of October. The key is to pay attention and observe the natural processes of the land that are visible and to be mindful of the Grandmother in the sky.

Also, it is important to remember that such names have changed over time. For instance, there was a time when Manoominikewi-giizis was not used as before and during the Anishinaabe migration from the East Coast. It was only after the Anishinaabe migration westward that manoomin (wild rice) was found in the estuary of Spirit Island near Duluth, Minnesota, and the practice of wild ricing was developed.

Last, depending on one's home region, there may be alternative names that some may utilize. In some regions Namebini-giizis ("The Suckerfish Moon") is used instead of Migiziwi-giizis, Onaabani-giizis ("The Hard Crust on the Snow Moon") is used instead of Aandego-giizis, Iskigamizigewi-giizis ("The Sap Boiling Moon") is used instead of Maango-giizis, and Waabigwani-giizis ("The Flowering Moon") is used instead of Zaagibagaawi-giizis, among others. The key is to listen to the language used in a particular region.

Did You Know?

The reflection of the sun upon the surface of the moon makes it the brightest and largest object that we see at night.

By affecting Earth's wobble on its axis, it helps with moderating a stable climate.

Every year, the moon moves approximately 3.8 centimeters away from Earth. Due to its having less gravity (because of its smaller mass), on the moon you would weigh 16.5 percent of your weight on Earth.

Because of its gravitational pull on the Earth (and the waters) against the Earth's rotation, the Earth's rotation has been very slightly and gradually slowing down – specifically 1.78 milliseconds over the last century.

Two-thirds to three-fourths of the effect on Earth's tides is caused by the moon's gravity. The remaining portion is the effect of the sun's gravity. From the human perspective, it appears that the waters are being pulled inward and outward from where we stand. What is happening is that the gravity of the moon and of the sun creates two bulges of water on opposite sides of the Earth. The Earth naturally rotates into (and out of) these bulges. No matter the phase of the moon, the effect on the tides is roughly the same. The higher tide takes place when there is a "full moon" and "new moon" because of gravitational assistance from the sun. This is because of the alignment, which is called syzygy. The lowest tide (called the Neap Tide) is caused by the moon and sun canceling out much of their gravitational effect.

It is thought that the moon may have formed due to a Mars-sized planetary body colliding with the early Earth approximately 4.425 billion years ago. This may have caused a debris ring around the Earth before condensing into a natural satellite we call the moon today, tilting the axis of the Earth and helping to create the seasons.

Gaagige-giizhig
The Universe

Every Indigenous culture worldwide has passed down teachings and stories of the beginning of life and origins of the universe. They may vary widely, but each acknowledges a process, purpose, and human beings being intimately connected to everything else in existence.

Anishinaabe teachings and stories about the process that brought the universe into existence are still held in great reverence. The language embedded in the name for the universe, Gaagige-giizhig, reveals that ancestors acknowledged the size, scope, and dynamic nature of the universe that surrounds us all. It literally means the everlasting sky.

It is said that the first sound that was made in the universe was that of a zhiishiigwan (a shaker).

Did You Know?

Evidence shows that our universe is roughly 13.8 billion years old. Not only is the universe continuing to expand outward, but it is also increasing in speed.

Consider the following. Add up all the grains of sand on all the beaches on Earth. There are still more stars in the universe than grains of sand on Earth.

The fastest spinning objects in the universe are called neutron stars, which emit a beam of radiation. As it spins, it somewhat resembles a lighthouse, giving off a pulse of light. This spinning motion gives the appearance of flickering light.

The multiverse is an idea that our universe is just one of many universes.

The initial "glow" of the universe that followed its beginning and rapid expansion can still be seen today as cosmic microwave radiation.

The farther away an object is, the longer its light takes to reach us. That means when you look out into deep space at distant stars, you are looking back in time.

You may feel that you are standing still, but there is more going on than you may realize. The earth you are standing upon is rotating at more than 1,600 kilometers per hour. The Earth is also going around our solar system (with the sun at the center) at more than 30 kilometers per second. Our solar system (moving within an arm of the spinning Milky Way galaxy) is moving at more than 828,000 kilometers per hour. The Milky Way galaxy is moving through the universe at more than 2.1 million kilometers per hour. The reason you don't feel it is similar to when you are riding in a car. If you stay at the same speed, the movement of the car isn't really felt. However, it is felt when the car increases or decreases in speed (or stops), which makes you move forward or backward. Since the motion of our planet and galaxy has been present and consistent since you were born, it isn't felt.

Anang
Star

Stars have a special place in Anishinaabe culture. As our ancestors looked up at the night sky, the star patterns and celestial events shaped the basis for many sacred stories that we still share today. Paying attention to events such as the locations of the stars (and their constellations) in the sky, the cycle of our moon, and seasonal changes in our natural environment, families better understood the natural calendar of the world. This helped generations survive and thrive.

Also shared across generations today are traditional and spiritual teachings, both connected to (and inspired by) the stars that help embed values which should ideally help Anishinaabeg maintain a good life and good relationships with one another.

One fundamental teaching is that the starlight is the thoughts of Gizhe-Manidoo (The Kind Spirit) being pulled back to their source of origin.

Anang – "Star"
Anangoog – "Stars"
Anangokaa. – "There is an abundance of stars."
Bagakaasigewag anangoog. – "The stars are shining brightly."

Did You Know?

We have a star very close to us. This is our Sun.

There are different kinds of stars, such as Red Dwarves, Blue Giants, and more. The names are determined by their color, size, density, and temperature.

Do you know how many stars there are?

There are approximately 200–400 billion stars in the Milky Way galaxy. There could be as many as 500 billion galaxies in the universe. That's at least 200,000,000,000,000,000,000,000 stars in the universe – some recent estimates and studies suggest more.

Beyond our Sun, the closest star to the Earth is Proxima Centauri. It is 4.2 light years away from Earth. This means it takes 4.2 years for light to leave Proxima Centauri, travel through space, and arrive at our eyes. This means when we look at Proxima Centauri, we don't see what it looks like at the moment we are looking at it. Rather, we see what it looked like 4.2 years ago.

Stephen's Quintet is a group of five galaxies, each with its own stars. They are visible in the constellation of Moonz, the Moose (commonly known as Pegasus). They are 290 million light years from Earth. This means if someone was looking at our Sun from one of these galaxies in Stephen's Quintet, they would not see you reading this book today. They would see when the sun warmed the dinosaurs on Earth. The light they would see would have left our Sun 290 million years ago.

Evidence shows a very special bond between us and the stars. The atoms that make up life on Earth and in our bodies can be traced to the cores of stars long ago. In the heat and pressure in early stars, elements such as carbon, nitrogen, oxygen, iron (and more) in our universe and in our bodies were cooked and created from lighter elements. As the stars approached the end of their lives, those stars collapsed and exploded in supernovae. This scattered those newly created elements across the universe, and they would, in turn, condense and collapse. This created new star systems. To create the amount of these elements in our universe, these events and explosions happened over and over. In time, this created more star systems, solar systems, and the ingredients for life itself. This is why every atom in our bodies (like all elements in the universe) can be traced to the cores of stars long ago.

Ojiig Anang
Fisher Star

If you were to watch the night sky for hours on end (or if you were to take a time-lapse video of the night sky), the stars would seem to rotate around a fixed point in the sky. The star closest to this point that doesn't appear to move is called Ojiig Anang – the Fisher Star.

The closest constellation outside of the space and pattern where the Ojiing Anang is located is Ojiig, the Fisher (also known as the Big Dipper). It is called the Ojiig Anang because Ojiig spends the years traveling around it endlessly.

During every season of the year, Ojiing Anang is visible and can be found in the northern direction of the sky.

Did You Know?

Ojiig Anang is also known as Polaris – the North Star.

During every hour of the night, year-round, the North Star is visible in the back of Maang the Loon (also known as the Little Dipper) and is the brightest star in this constellation.

Look north and you will find Polaris.

If you travel north on the Earth, it will appear higher in the night sky.

As you begin to travel south on the Earth, it will appear closer to the horizon.

Also, if you were standing at the North Pole, it would appear directly overhead.

A spot in the sky that aligns with the axis of the Earth (and appears to us as a fixed point in the sky) is called a celestial pole.

Polaris is very close to this point, and it appears not to move very much from this point while all other stars in the sky move around it.

Polaris is a variation of "Stella Polaris," a name given in the Renaissance era, meaning "polar star" in New Latin.

Many names, such as Dhuva (in the Hindu Puranas) and Stella Maris, the Star of the Sea (in thirteenth-century France), have also been given to Polaris.

Polaris is a triple star system of Polaris Aa and Polaris Ab (respectively a primary star and a yellow supergiant) in orbit with another star, Polaris B.

Polaris has not always been the star at the celestial pole and will not be in the future. As the Earth rotates, there is a slight natural wobble. This is called precession. It creates a slow change in the rotational axis of the Earth (which causes a slow change in the celestial pole from our vantage point). This slow process takes 25,772 years to complete fully and begin again.

This means that although Polaris (Ojiig Anang) is our current North Star, the point of the sky that the stars appear to rotate around changes over time. Thirteen thousand years ago, Vega (Gichi-Ogimaa Anang) was our North Star and will once again be so in roughly 13,000 years. Although several other stars (such as Thuban, Fawaris, Errai) have been at this place in the sky due to precession, Vega has been (and will be) the brightest.

Ojiig
The Big Dipper

All throughout the year, Ojiig can be seen in the northern sky.

He can be seen highest in the sky during the spring and close to the horizon in the fall.

During the evenings of spring and summer, he shines brightly and makes himself clearly visible.

It is said that a long time ago, the birds that accompany the return of the spring season and scatter life didn't arrive as they had been held in another realm by a being. Enveloped in thick clouds, the birds were disoriented and could not find their way back. Many animals attempted to climb a tall cedar to an opening in the sky, but only the determined Ojiig (fisher) could accomplish this feat. He spun his characteristic tail to disperse the clouds. As the once-disoriented birds got their bearings, Ojiig directed them back to the hole so they could return home. The last one to leap from the realm was Ojiig, who was struck in the tail by an arrow shot by the angered being for releasing and guiding the birds back home.

Because the fisher made a great sacrifice to help the birds (who help to bring the regrowth of spring) return home, he is visible throughout the year in an honored place in the sky.

The stars in this constellation make up the body of Ojiig and the arrow pinning the tip of his tail to the sky.

In some regions of Anishinaabe territory, it is said that (rather than being pinned by the tail), he was made into stars and placed up into the sky as an eternal reminder of his sacrifice.

Did You Know?

The Big Dipper (or Ursa Major) has captured the imagination of countless generations of people across the globe. Cultures from Greece, Ireland, Germany, Romania, Hungary, Vietnam, China, Japan (and many more) each have stories and teachings related to this constellation. Each story is different and envisions the star patterns in different ways, including a great bear, a ladle, a crustacean, a plough, a wagon, and many other beings and objects central to stories.

The stars of this constellation have not only helped in storytelling and passing cultural teachings across generations, but they have also aided explorers and travelers with navigation.

Because of its prominence in the sky, this notable star pattern has been used by amateur and professional astronomers to locate other stars in the sky.

The constellation is a wonder that crosses cultures, continents, and generations.

Because it can be seen year-round, it is said to be circumpolar.

Gichi-Ogimaa Anang
Vega

Imagine yourself living 13,000 years ago.

As daylight comes to an end in the evening, there is a special star that sits high in the night sky.

Throughout the nighttime hours, all other stars seem to slowly move about the sky. However, that one special luminous blue-white star shines near a point of the sky that remains fixed.

If you were to take a time-lapse video over an extended period, you would see the other stars move around this fixed area.

By its starlight, it reminds you of the northern direction for travel, for ceremony, and to help you return home.

It is consistent, supportive when needed, and helps to lead you where you want to be in this beautiful land in the North.

There will be a day when it will not be the fixed point in the sky, but rather the fixed point will be called Polaris.

However, the former star will be the fixed point again far into the future.

This is Gichi-Ogimaa Anang.

The Great Leader Star.

Did You Know?

Vega is the fifth-brightest sun in the night sky and is a blue-white color.

When looking high above in the sky toward the east/northeast, it can be seen in the evening during summer months and is most visible in (and around) the month of June.

Vega is sometimes known as the Harp Star because of its location in the constellation of Lyra – the Harp. It is the most visible star.

Locating the imaginary "summer triangle" (made up of Vega, Deneb in the Cygnus constellation, and Altair) can help stargazers locate Lyra.

It's difficult to speak about Vega and not discuss the cycle of precession.

Each night, pay attention. The stars appear to change position. However, some points remain fixed because they sit along the Earth's axis. These are called celestial poles.

However, the Earth has a slight wobble like the spinning of a top.

This is called precession.

During a cycle of precession, the north and south celestial poles (the two imaginary points in the sky that remain fixed in the sky as everything appears to rotate around them) slowly change. The celestial poles seem to move in circles, finishing a cycle every 25,772 years.

Today, Ojiig Anang (Polaris) is the star that sits at Earth's north celestial pole. This is why it is called the North Star. This wasn't always the case – nor will it be the case far into the future.

At roughly 12,000 BC, the pole was only five degrees from Vega and was the brightest pole star. It will be a pole star again in AD 14,000.

Gaa-bibooniked
The Wintermaker

Each year, there comes a time when the season of autumn gradually shifts to winter. It begins with a chill in the air. Birds begin to make their way south to warmer temperatures. Finally, the chill turns to flakes of snow blowing across the land and the waters.

Ancient teachings and stories speak of a being called the Wintermaker who helps usher in the freezing temperatures and the cold weather. Bringing his extended arm across the horizon, he pulls the frigid air and strong breeze from the north and pushes them to the south.

There is no question in the villages, communities, towns, and cities that are in the path of the freezing temperatures brought forth by the Wintermaker that the new season is on its way.

Gaa-biboonikéd (also called Biboonikwewinini to the south) is visible toward the southwest from November until March, with increased clarity occurring in January. Before the season begins to change to winter, he appears in the sky. One can see his outstretched arm, pulling the icy temperatures for winter to arrive.

Did You Know?

The constellation Gaa-biboonikéd also carries the name of Orion from Greek mythology. It is one of the most recognizable constellations in the night sky.

One of the most recognizable features of Orion, the Great Hunter, is Orion's Belt. It is made up of three bright stars known as Alnitak, Alnilan, and Mintaka. Alnitak is more than 100,000 times more luminous than the sun in our solar system.

Orion has served explorers, sailors, and travelers throughout history as an effective navigational aid, as well.

Cultures from around the world have told stories and passed down teachings connected to this constellation.

In Babylonia, Orion was known as the Heavenly Shepard. In Armenia, he is known as Hayk. In Ancient Egypt, the god Sah. In medieval cultures in the Middle East, Orion was known as the Giant, "Al-Jabbar." In Siberia, Orion is also seen as a hunter. The depiction of a great hunter is shared in ancient Hungry under the name Nimrod.

Orion remains a popular feature in the sky. It is one of the first constellations amateur astronomers are taught to locate.

Maang
The Little Dipper

For countless generations, Maang (the Loon) has been a highly respected living being to the Anishinaabeg.

Those born into the Loon Clan are born with the responsibilities of leadership internal to the nation. Countless Anishinaabe civil leaders have come from the Loon Clan.

Just as Maang rests in the night sky outlined in stars, loons have a special relationship to the night sky. This connection is patterned on their backs, resembling countless white stars on a dark sky.

When you see loons on our lakes, take a moment to appreciate the white pattern around their necks resembling a necklace adorned with Miigis shells – the shells sacred to the Anishinaabeg.

Within the body of Maang is Ojiig Anang, the Fisher Star (also called Polaris or the North Star).

In the Northern Hemisphere, Maang can be seen at all hours in the northern sky throughout the entire year.

Maang is inverted in comparison to Ojiig.

Did You Know?

In English, Maang is called the Little Dipper.

It is often regarded as accompanying the Big Dipper (Ojiig) in the sky.

Teachings and stories throughout history have regarded this constellation as notable in the sky.

In Babylonia, it was known as a wagon in the heavens.

Stories and teachings about deities, canines, wolves, cows, bears, even trails of fire were told by many civilizations throughout history in places.

This constellation has historically been crucial for sailing and navigation because the brightest star in this constellation is Polaris – the North Star.

Although this constellation can be seen throughout the year, the early nighttime hours during the month of June present clear viewing. This is because only three of the Little Dipper's seven stars (Polaris, Kochab, and Pherkad) are noticeably bright.

One effective way to locate the North Star (which is housed in the Little Dipper constellation) is to find the Big Dipper. A line can be drawn between the two stars that make up the front of the Big Dipper (beginning with Merak and to Dubhe). If this line continues through Dubhe and further into the night sky, it will point directly to the North Star.

Bagonegiizhig
The Pleiades

Bagonegiizhig is a prominent feature visible between the months of October and April (most clearly in November) and visible from the far Northern Hemisphere to the far Southern Hemisphere.

It is widely known as a signal that winter is coming and remains in the sky from late fall to late winter.

Bagonegiizhig translates to "the hole in the sky" (or in some regions, "the hole in the day").

According to teachings along the Rainy River (and in many other regions), this ring of stars is very important and connected to Anishinaabeg as a place of origin, the location where Gizhe-Manidoo came through to provide important teachings, and the location that is traveled toward (and entered) to begin four days of travel when our lives on Earth are completed.

It is shared that although only seven stars are easily visible, there are four rings of seven/eight stars.

However, they are difficult to perceive because the vantage point from Earth is not perfect.

It is also shared across generations in regions such as the Rainy River that the seven stars are the location of the poles of one of the most sacred ceremonies to the Anishinaabeg – the Jiisakaan (or shaking tent).

In many other regions, some speak of the seven stars as seven Grandfather stones from a sweatlodge.

The concept of Bagonegiizhig is so highly respected, there have been multiple Anishinaabeg who have carried this name. Bagonegiizhig (Hole-in-the-Day the Younger) was an influential leader.

To find Bagonegiizhig in the night sky, locate Aadwaa'amoog. Connect the three stars by an imaginary line, continue to the right of the sky, and you will naturally find Bagonegiizhig.

At times, it is difficult to see when looking directly at the cluster. Looking near its precise location in the sky can help.

Did You Know?

Present in the northwest of the constellation of Taurus, the Pleiades remains perhaps the best-known open star cluster and has fascinated cultures across the globe for countless generations.

It is commonly referred to as the Seven Sisters.

This is because six easily visible stars and one star (which is a bit more difficult to see) have been acknowledged the world over since ancient times.

Among the Greeks, stories were connected to these stars – seven divine sisters known as Taygeta, Celaeno, Merope, Electra, Aleyone, Maia, and Asterope.

On a very dark and clear night, nine stars are visible.

Among the Ancient Greeks, stars eight and nine were their father and mother – Atlas and Pleione.

The ancient Maya, the Chinese, the Japanese, the Māori, Indigenous Peoples in Australia, the Celts, Indigenous Peoples of Hawaii, Babylonians, and the Oceti Sakowin are just a few cultures across time and the globe who have also connected cultural teachings and stories about the Pleiades.

In some locations, up to fourteen stars can be seen when away from light pollution from cities.

In recent times, the automobile company Subaru depicts the six most visible stars in this cluster as its logo.

With the improvement of telescope technology, we know now that there are not only six, seven, nine, or fourteen stars. Rather, there are hundreds of stars present among visible interstellar gas clouds passing through the Pleiades.

Because they are blue giants (hotter than our own star – the Sun), the clouds glow from the stars' radiation.

Moonz
Pegasus

Moonz is the large and respected animal that calls the Northern Hemisphere home.

The name among related peoples of the Algonquian language family is the etymological origin of Moose.

It is a clan among the Ojibwe-Anishinaabe Hoof Clan family. The Hoof Clan (adorning Clan colors of brown and black) are responsible for keeping the artistic traditions of the Anishinaabeg alive, the consistent use of gentle words, poetry, communication, peacemaking, reconciliation, community conflict resolution, and housing. Moonz is also at the center of several traditional stories and legends.

Still a popular and very nutritious game meat, moose have been hunted for survival for countless generations. To this day, grateful hunters share the meat of their most recent hunt with fellow family and community members.

Long ago, survival depended on the practice of using every part of the animal for sustenance, warmth, and tools. This is still done by many today. Similar to deer, caribou, and buffalo, hides from moose are still utilized to make dewe'iganag (drums).

A common practice by hunters of generations long ago and today is to show the utmost respect to the moose when harvested. The dewlap (or bell) that hangs near the chin of the moose is removed, filled with asemaa (tobacco) and gratitude, and hung in a tree at roughly the height of where it hung when the moose was filled with life and walked upright.

Visible from late summer until mid-winter in the Northern Hemisphere, Moonz can be seen in the night sky.

Moonz can be seen by inverting the constellation of Pegasus.

It is huge – the seventh largest constellation in the sky.

Not only visible in the sky by means of starlight, moose also appear on a variety of rock paintings through the Great Lakes and Lake of the Woods regions. There are always important connections between the two.

Did You Know?

The mythical winged horse Pegasus is one of the notable constellations created by the astronomer Ptolemy who lived until AD 168.

Although thirteen very visible stars are connected to create the constellation of Pegasus, satellite imagery reveals more than 2,000 among them.

The brightest star in the constellation is Markaband.

The Great Square of Pegasus creates the main body of the winged horse and is made by connecting Markab, Scheat, Algenib, and Alpheratz.

While the Ancient Greeks identified this constellation as a winged horse, countless cultures across the world and throughout history have shared different stories and teachings.

For instance, stories of a giant tortoise were told in Ancient China and are still passed down the generations by some today.

Onwaajige Anang
Halley's Comet

For generations, many Anishinaabeg have turned their eyes to the stars and observed patterns over time.

A long-tailed celestial light would arrive, create amazement, depart, and return many years later.

Over the generations, many shared that this amazing celestial being regularly returned to observe the Anishinaabeg to ensure respect, doing things in a good way, and fulfilling responsibilities were continuing and vibrant.

This event also served as a reminder that those who helped in a divine way were always present.

Because of the connection to many important teachings (as well as the expectation of a return regularly fulfilled), Onwaajige Anang is the name given to this celestial body.

Onwaajige Anang – the Star of Prophecy.

In some regions, the name of this celestial body describes a long, shining tail during travel across the sky.

Did You Know?

Halley's Comet is named after astronomer and mathematician Edmond Halley, who studied and understood that a variety of comet sightings throughout history were the same comet's reappearance (and could predict when it would return).

Halley's Comet is truly a sight to see. It is visible from the Earth every 75 to 76 years.

With a very elliptical orbit, it is the only comet that can be seen by the naked eye that one can see twice in a lifetime. With its relatively short orbit (less than 200 years), it is classified as a short-period (or periodic) comet.

This orbit brings Halley's Comet continuously through the solar system, closer to the Sun and then further away.

Gravity is the key.

As the comet gets closer to the sun, a long tail becomes visible due to evaporation of ice and sublimation of its material composition, release of gases (called outgassing), the creation of an atmosphere from this process (called a coma), and absorption of solar light (called fluorescence).

However, as its orbit takes it further away from the sun into the far reaches of space, a process of cooling and freezing occurs.

Ancient cultures throughout the world have witnessed Halley's Comet and connected many stories and teachings to its appearance in the skies.

Indigenous Peoples across the globe continue to share stories of Halley's Comet – ranging from celestial beings in travel, the smoke of a village fire, shining spears being thrown across the expanse of the sky, a long serpent, a sign of coming drought or sickness, and more.

The Ancient Greeks and Romans placed great importance on its arrival as either a positive or negative divine omen.

At times, it was connected by some with coming destruction, conquests of kingdoms, or natural disasters across Europe and the Middle East.

During the time of the Han Dynasty in China, records describe a broom-like star in the skies.

The last time Halley's Comet was visible from the Earth was in 1986 and it will return in 2061.

Madoodiswan
Corona Borealis

The madoodiswan (sweatlodge) is a womb, a place of healing, a place of ceremony, and a source of insight.

Pronounced as "midoodoosawin" by several Anishinaabemowin speakers carrying ancient linguistics, the name of a sweatlodge expresses "carrying the life energy of the universe of my Mother."

It is one of many manidookewinan (ceremonies) carried by many Anishinaabeg with great respect, responsibility, and care.

Madoodiswan contains seven stars positioned in a semicircle. The highest point in this curve is the top of the lodge.

The seven stars contain many teachings, including the significance of the number of stones commonly brought into a lodge, as well as a very in-depth understanding of the spiritual structure above and below the Earth.

It rests in the Northern Celestial Hemisphere and can be seen during the spring and summer.

Spring is a time of rebirth, while summer is a time of gathering. Although ceremonies take place all throughout the year, summer is very special. As families returned from winters in the bush and completed annual maple sapping, friends and families (who may not have seen one another for months) would gather in great numbers. Along the Rainy River, massive gatherings and ceremonies would take place in locations such as Ginwaajiwanaang (Long Sault Rapids) and Gojijiing (commonly referred to as Pither's Point, next to Baawitigoonsing Neyaashiing) during the summer season.

Madoodiswan is an inversion of the Corona Borealis.

As it can appear fairly dim, it is best to view the lodge on very dark, clear nights.

Did You Know?

The Corona Borealis (also called the Northern Crown) contains a series of seven luminous stars that resemble a crown in the sky.

The seven stars are each a tip of the crown.

They are known as Alphecca, Nusakan, Gamma Coronae Borealis, Epsilon Coronae Borealis, Theta Coronae Borealis, Delta Coronae Borealis, and Iota Coronae Borealis – the brightest of which is Alphecca.

Many stories and teachings have been connected to Corona Borealis throughout history.

Stories of a crown gifted to the daughter of Minos by Dionysis, a silver castle, a garden with flowers, a fractured dish, a hole through which smoke can carry prayers, a den of a bear, a nest, a gathering of sisters, and a council of Elders are just a few spanning across generations, cultures, and locations.

Ma'iingan
Canis Major

Ma'iingan (the Wolf) holds a very special place in Anishinaabe stories.

Not only do stories tell of him being the first companion of the original human spirit (who at the time did not have a name), but the destinies between wolf and humans were to be interlocked for all time after they were told to go their separate ways. What happens to one will happen to the other.

The descendants of wolves are dogs, and there are many stories among the Anishinaabe as to why they are so special, respected, and share such a great bond with humans.

Ma'iinganag mate for life, are very community-minded, communicate and strategize well, and hold a special place in teachings and culture.

Wolves are so influential; they can massively change and even improve entire ecosystems in a particular area.

Did You Know?

From December through March, the constellation Canis Major can be seen by those who call the Northern Hemisphere home when looking toward the South.

Stories among the Ancient Greeks tell of a pair of sled dogs belonging to Orion who assist in his hunting voyages. The name Canis Major refers to the larger of the two, and both are stellar neighbors to Orion.

Canis Major contains ten primary stars that outline the shape of a dog, but only eight are easy to see.

One of the stars in this constellation is Sirius – also called the Dog Star. This star gets its name (which expresses something scorching hot) because it is the brightest star in the night sky.

Sirius rests in the chest of Canis Major.

The brilliance of Sirius coming up over the horizon at sunrise once a year was so significant in Ancient Egypt, calendars would begin their cycle with this event.

Among many in India, Canis Major is a hunter, and Orion is a deer that is being pursued.

Nanaboozhoo
Scorpius

Nanaboozhoo is a central being in stories that Anishinaabe tell with great reverence.

At times, the names Nanabush or Wenaboozhoo are used in some regions.

A teacher (and metahero embedded into cultural stories), he is referred to as half-human and half-spirit. As the human side makes mistakes common to the human family, the spirit learns. As stories are told (often during the winter months), future generations can be the recipients of many lifetimes of those who came before, who watched, listened, paid attention, and interwove ethics and solutions to common problems into stories.

Ancient understanding of the language connects these names.

Nanaboozhoo – using one's breath of life to speak very clearly about the experience of life.

Wenaboozhoo – using one's breath of life to ask very clearly about the experience of life.

It is said by those who have received this understanding that whether one is speaking clearly about (or asking very clearly about) life, it is either Nanaboozhoo or Wenaboozhoo speaking through them.

South of 40 degrees latitude in the Northern Hemisphere, Nanaboozhoo can be seen sitting low in the sky throughout the summer for those viewing the sky. He can be seen toward the southern sky.

Kneeling and pulling his bow back, it is said that he is quietly hunting. Depending on the region, Nanaboozhoo is said to be hunting a variety of animals for a number of reasons. The most common of these animals is the spirit of Mishibizhiw, the underwater panther.

Did You Know?

Scorpius is mostly viewed in the Southern Hemisphere while sitting high in the northern sky. You can see this constellation until the end of September.

However, if you are in the Northern Hemisphere, look low in the southern sky near the horizon until the middle of September.

Around 10:00 p.m. is a prime time to spot Scorpius.

There are eighteen main stars in the shape of a fish-hook or a curved scorpion's tail.

The brightest star, Antares, refers to a rival of the Greek deity Ares, the God of War.

It was said that Orion and Scorpius are arch enemies in stories passed down generations. Pridefully, Orion bragged about his hunting skills, and Scorpius was sent to take his life. To ensure they would not cross each other's paths, they were placed far apart in the sky.

One appears while another disappears.

Cultures all over the world describe this connection of stars in different ways.

Some say it is a hunter, a falcon, a crocodile, or a fish-hook belonging to a demi-deity named Maui.

Waaban Anang
The Morning Star

Throughout the generations, Anishinaabeg have gazed in wonder at the process of sunrise. Before the sun breaks the horizon, the sky begins to lighten. Even before the eastern sky takes on this visual beauty, a celestial body shines like a beacon from the east at many times of the year.

This is known as Waaban Anang – the Morning Star.

Stories passed down the generations often tell of a beautiful lodge to the East belonging to a beautiful (and very important) figure connected to the being who was named Anishinaabe that announces the dawn and rising of the sun.

Did You Know?

The celestial body that is visible in the eastern sky before sunrise is actually the planet Venus.

Venus has been identified by many cultures (both ancient and contemporary) with a name approximating a morning star.

It is not actually a star, as planets do not give off their own light.

However, Venus can appear as a star and shine intensely brightly because of the clouds that surround the planet. As it is the second planet closest to the sun, the solar light is brilliantly reflected.

Because of Venus's position in the solar system, those on Earth cannot see it in the middle of the night (except during very specific times due to an elongated orbit). However, just before sunrise is a notable time when it can be visible. Hence, a morning star.

Many cultures throughout history have connected teachings and stories to Venus's appearance as a morning star. The Ancient Egyptians shared stories of Tioumoutiri while the Ancient Greeks described this morning "star" as a bringer of light named Phosphoros. Across history and tradition, stories from the Mayans, Maasai, Vietnamese, and more attached stories and teachings to this wonder in the morning sky.

But that's not all by a long shot. More information about Venus and its place in culture and history will be further explored below.

Biidaaban, Waaban, Zaagajiwens, & Mooka'am
The Process of Sunrise

Just like countless generations of the past, Anishinaabeg close their eyes each night and hope for enjoyable and insightful dreams. There comes a time as night ends and a new day begins.

When a new day begins, it is important to remember that it is not a single moment, but a long process that repeats each morning.

Before the sun rises, the night sky begins to become lighter and slowly becomes brighter.

This process of moments is called **Biidaaban** – "The dawn is coming."

The moment of daybreak occurs, which is called **Waaban** – "It is daybreak."

The brilliant sunlight pours over the horizon, a flow of light and energy that is called **Zaagajiwe** – "The light is moving out over the horizon."

Finally, the Sun begins to climb the sky to warm all beings with light.

This action and motion of the Sun is called **Mooka'am**.

As Gimishoomisinaan Giizis (Our Grandfather Sun) is perceived as a living being, this literally means "He is getting up" or "He is rising."

This description perfectly reflects what is captured in the concept of a sunrise.

It is said that time stops for a moment when the sun breaks the horizon each morning. According to the Anishinaabe stories, this is when Migizi (the bald eagle) prevented the Sun from being blown out by a being by calling out four times and pleading for the continuation of humankind long ago. This underscores why each day is a gift and why sunrise ceremonies are done with great respect.

This process (beginning with the appearance of the Waaban Anang, depending on the time of year) is spoken about with great reverence. There are many stories carried within lodges, as well as teachings that are connected to each moment of this process (and the language that defines it) that brings a new day.

Did You Know?

Living beings that call the Earth home often perceive that the sun is rising in the East.

If one looks from an astronomical perspective, the Earth is consistently rotating on its axis. As the Earth rotates due east, the Sun gradually is revealed and becomes visible. As the Earth continues to rotate, beings on the Earth rotate under and away from the Sun. As we do not feel the Earth rotating, it is easy to perceive the sun rising, traveling the length of the sky, and lowering in the West.

Did you know that sunlight can be seen lighting the sky (and the sun is visible) even before it breaks the horizon? This is due to something called refraction. Refraction can be thought of as the bending of light when moving from one substance to another, the density of those substances, and the angle at which the light arrives. If light hits the atmosphere at a shallow angle rather than head-on, it bends toward the Earth. Therefore, the sunlight and the sun can be seen from our vantage point even before the moment of sunrise. This effect also happens at sunset ("Bangishimon") when the sun appears to lower down and set ("Bangishimo") and still remain visible for a time after it is behind the horizon from our vantage point.

As the Sun has been revered as a life-giver by countless cultures across time and geography, many stories and teachings from across the world connect to our nearest star.

In Greece, the deity Helios appeared in many stories as he began to climb the sky from the horizon at sunrise in his fiery chariot and move to the West throughout the day. He continued to do this every day.

Stories from Egypt featuring Ra taking a boat across the sky each day, beginning at sunrise, were passed down through generations.

Stories from Norse cultures spoke of four horses pulling a chariot from the East (at sunrise) to the West.

Mishibizhiw
Leo, Cancer, and Hydra

If someone living in the Northern Hemisphere looks to the sky during late winter and spring, they will see a very ancient and notable being revered by the Anishinaabeg outlined in the stars above.

This being is known as Mishibizhiw – the great underwater cat.

A very large being often depicted and spoken of as an underwater panther, this powerful being is both respected and honored.

One of the many roles and responsibilities of Mizhibizhiw is that of a protector.

Protecting all that occurs below the surface of the water, Mishibizhiw is said to be part of a universal pair of forces existing in the world – the other being Binesi, the Thunderbird, who watches over (and protects) all above the surface of the water.

Sacred stories transmitted down many generations firmly establish that though complementary beings, Mizhibizhiw and Binesi often come into conflict with one another.

Calling sacred expanses of water, sets of rapids, and locations of fast-flowing currents home, very special offerings of cloth, food, and tobacco are still regularly given to acknowledge and show respect to Mishibizhiw.

It is not uncommon for those who harvest fish from lakes and rivers (as well as those who travel by means of the water) to acknowledge Mishibizhiw with a tobacco offering for a safe journey and to ensure they are utilizing his home in a good way.

When standing in the Earth's Northern Hemisphere, Mishibizhiw can be seen during the spring in the northern sky – clearly during April and May from early evening and throughout the night.

A large, curled tail is Mishibizhiw's characteristic feature, which is described in teachings and stories (and visually expressed among various rock paintings); it can be found as the curve connecting the neck and head of the more mainstream constellation of Leo. Mishibizhiw's back contains some stars from the constellation of Cancer, and his face is defined by the head of the constellation of Hydra.

Did You Know?

Leo, Cancer, and Hydra are three constellations that remain well-known to stargazers.

Leo is Latin for Lion.

To the Ancient Greeks, stories and teachings told of a Lion that was battled and killed by Hercules.

Seen by many as a shape akin to a question mark, the mane of Leo the Lion is often compared to a sickle in the skies.

In stories and teachings of the greater Mesopotamian region connected to this constellation (such as in Babylonia, Sumer, and among the Turks, Jewish tribes, and Persians), the stellar being of a Lion was common.

Cancer is Latin for Crab.

To the Ancient Greeks, stories of a crab who bit Hercules on the foot were told for countless generations. According to these stories, Hercules took its life, and it was placed among the stars by a notable goddess.

The poet Dante referred to this constellation in his written work *Paradiso*, as well.

It is visible from the Northern Hemisphere near the constellation of Leo in the early spring.

The great sea serpent that Hercules battled during his journey is also visualized in the sky by the constellation of Hydra.

One of the largest star patterns in the sky, many cultures throughout history attributed cultural stories and teachings to the stars they observed.

Among many Arabs, the Hydra's head was described as a place of non-inhabitation in the sky.

Within Hindu teachings, Hydra was described as a coiled serpent (as well as stars in relationship to the Azure Dragon and the Vermilion Bird to the Chinese).

Hydra can be observed in the Northern Hemisphere between January and May while viewing the southern sky.

Gaa-madoodood
Hercules

Connected visually (and within stories) to the madoodiswan (sweatlodge) in the starry sky, Gaa-madoodood can be understood as "the one who is attending the sweatlodge ceremony."

Across other regions, Gaa-madoodood has also been called many names by Anishinaabeg. Although greatly varied, these names are all defined by a person choosing to attend a sweatlodge ceremony for healing and being very tired afterward – but healed in body, mind, heart, and spirit.

Anishinaabe teachings connected to Gaa-madoodood share that when attending such a lodge (or any ceremony), it is crucial to approach it with great respect. It is also important to seek out not only a gifted individual who carries the lodge and responsibilities of it, but someone very respectful and respectable.

Many of the most notable spiritual leaders in history have clearly stated that Anishinaabemowin is not only helpful, but absolutely necessary for a vast majority of ceremonies to be conducted. It brings them to life, fundamentally activates them, and there are things that can only be expressed (or understood) in Anishinaabemowin.

Did You Know?

Hercules is a central character of Roman stories and mythology who was adapted from the Greek hero Heracles.

Attributing stars to a man kneeling, this particular pattern of stars resembles an angled square with bent legs, a four-sided torso, a head, and arms bent and lifted upward.

Various other cultures, such as Indigenous Australians, the Chinese, and those across Mesopotamia, also perceived this constellation as a being among the stars with a diversity of connected stories and teachings.

Often associated with the stories of Gilgamesh among ancient Sumerians, it remains a notable constellation across the generations.

If you are standing in the Northern Hemisphere, look to the northern sky during the summer months.

Bangishin Anang
Falling Star

Although not a constellation, a falling star remains one of the most awe-inspiring sights to the human family – no matter the cultural stories and teachings connected to the event.

Like all branches of the human family, generations of eyes across regions of Anishinaabeg looked to the heavens when an illuminated pulse and flash pierced through the night sky.

This continues today.

Like a ball of fire, this event has brought wonder, inspiration, and (at times) fear to those who stand in amazement and observation.

Bangishin Anang, literally "The star is falling," expresses this sight.

Banishinoog Anang, "The stars are falling," conveys that multiple stars are falling to the Earth.

Gwiingwa'aage (wolverine) is a being that ancient Anishinaabe stories acknowledge as having originated on Earth by means of a falling star or meteor. Gwiingwa'aage contains sounds that paint a picture of light shooting and arching across the sky and falling to Earth.

Did You Know?

The luminous streaks of light that have captivated humanity are not really stars (as the descriptions of falling stars or shooting stars suggest). They are known as meteors.

A meteoroid is a celestial body moving through the solar system – such as rock – and can be very small to large, such as an asteroid.

If this mass enters the Earth's atmosphere, it begins to burn up and is known as a meteor.

As it moves through the atmosphere, strong resistance and drag begin acting on the object, and it begins to burn.

Long streaks of light are left as the meteor burns up on this trip through the atmosphere.

However, they are not rare. It is estimated that roughly 44,000 kilograms of these types of objects fall toward the Earth on any given day.

A vast majority burn up in the atmosphere.

There are events where many occur in a short period of time. These are called meteor showers, and astronomers have developed ways to predict when they will happen.

Binesi
Cygnus

Among Anishinaabe teachings and stories, there is one supernatural being that (outside of the Creator) carries the energy of both the man and the woman – fire in the form of the lightning it brings and water in the form of rain.

This being is Binesi, the Thunderbird.

The name Binesi can refer to the spiritual being of the Thunderbird or a particular bird of prey that is its physical representation in our world.

In many regions, the name Animikii is used to refer to the spiritual being of the Thunderbird or that family of beings known as thunderers.

The distinctions and commonalities between Binesiwag and Animikiig (within crucial linguistics, stories, and teachings) are still carried by many today.

Protecting all life over the surfaces of the water, the Thunderbird is one of the most revered beings throughout Anishinaabe history, stories, and teachings.

Complementary to (and at times in epic conflict with) Mishibizhiw, one of the most important roles of the Thunderbird is protection and maintaining balance.

Another defining responsibility of the Thunderbird is that of restoration. It is not uncommon for storms to roll in across the land, and lightning may strike the land. By means of these lightning strikes, large fires are often ignited. As the landscape burns, both the long-standing species of plant life and invasive species that may be damaging are ignited and burned away. However, it is the long-standing species that grow back.

This is very similar to purposeful seasonal burns done by Anishinaabeg at Ginwaajiwanaang, near Manidoo Baawitigong, and many other places.

Anishinaabe reverence for the Thunderbird is often expressed in symbology used by the Anishinaabe Nation.

Binesi can be seen in the starry skies during the months of summer and autumn by looking to the northern skies.

It is commonly referred to as Cygnus by mainstream stargazers. However, Binesi is Cygnus inverted. The head of Cygnus is the bottom end of Binesi.

In some regions, this constellation of stars is referred to as Ajijaak (the Sandhill Crane). According to many stories and teachings, Ajijaak has a special role among the Anishinaabeg as one of the prominent civic leadership clans and a being who assists souls by helping bring them into the world.

In other regions, the inverted Binesi is perceived as Waabizii – the swan.

Did You Know?

The constellation Cygnus is well-known to contemporary astronomers and stargazers throughout history and in a variety of locations.

Visualized by many as a cross pattern in the sky, Cygnus the Swan's lengthy neck emerges from a short body with wings on the side.

Stories among the Ancient Greeks told of Zeus shapeshifting to the form of a beautiful swan in order to seduce a Spartan queen. In some versions of the story, the focus of Zeus's attention was a being named Nemesis.

Another story is attributed to the character of Orpheus.

The Chinese and the Hindu perceive Cygnus as a bridge of a celestial river or as connected to an important period of the morning hours, respectively.

One of the stars found in Cygnus (Deneb) roughly translates to "a hen's tail" in Arabic.

Mishiginebig
Draco

Along the Rainy River, Manitou Rapids, and countless locations across the region, a very respected being has been acknowledged across the generations.

It is a helper, a being that helps remind Anishinaabeg to hold fast to their responsibilities, and a majestic creature that many continue to pass on sacred teachings about (and personal accounts of witnessing them making themselves visible) to those still yet to be born.

Mishiginebig, commonly referred to as the Great Snake, is a noted helper within stories and teachings who provides direction and assistance.

Offerings of tobacco, cloth, and food are still put out to show gratitude to Mishiginebigoog, who use lakes, river systems, and travel routes unseen to those that walk upon the Earth.

When making themselves visible, they bring attention and guidance to those to whom they reveal themselves.

It is important to remember that ginebigoog (snakes) are not adversaries to human beings. They are respected and feared by some, but they are, most importantly, helpers with intimate knowledge of plant life and locations of medicines.

Mishiginebig can be seen in the northern sky during the summer months and is clearest in July.

Through the generations, some stargazers refer to this constellation as Draco.

Did You Know?

In Latin, draco refers to a dragon.

Containing seventeen named stars, Draco is connected to the story of Hercules as Ladon the Dragon, who once guarded golden apples in a very beautiful apple tree surrounded by gardens.

Its Arabic name refers to a great serpent, as well.

For many stargazers in the Northern Hemisphere, Draco never sets below the horizon. This is called a circumpolar constellation.

Ikwe Anang
Venus

Both men and women are acknowledged in balance with one another in Anishinaabe culture. This is demonstrated clearly in the linguistics of Anishinaabemowin, as well as within ancient symbology, ceremony protocols, and within traditional stories. The balance and equality between (and significance of both) men and women is consistent and profoundly demonstrated by the responsibilities of many women as midwives and by many men who officiate ceremonies at the end of life – one helps to invite the spirit into the world while the other helps to send them home. This balance can also be seen in the creation of life by means of the union between Anishinaabe and the Firekeeper's daughter and their four sons, who forged four more unions with four women respectively in the four directions.

Although balance and equality between men and women are firmly demonstrated in the language, protocols, stories, and symbology, it is also acknowledged that each brings something unique to the world.

When looking up at the sky, there is a reminder of one of the gifts that women possess in the world. This gift is giving birth. Although men and women create life together (and are life-givers by means of their union), women carry the baby through its development in the womb until birth.

Ikwe Anang (or the Woman Star) is a brilliant sight to see – a luminous celestial body that rests in the eastern sky. It is very visible during the morning hours.

However, after appearing in the eastern sky for roughly nine months, it disappears and emerges again in another location of the sky for another nine months before this stellar cycle repeats.

Because of the nine months of observation, stories and teachings connect the roughly nine-month gestation of a baby to this celestial event.

In some regions, the appearance of this star in the western sky during the evening is acknowledged as a man star, recognizing this sacred balance.

Did You Know?

Venus is a unique planet.

Due to its position in the solar system (closer to the Sun than Earth) and its orbit, something very special happens when viewing Venus from Earth.

For a nine-month period, it is visible in the eastern sky during the morning hours. It will then disappear before emerging visible in the western sky as an observable object in the evening.

Because of this event, the planet Venus has been consistently connected to the concept of a woman. It is no wonder that Venus is named after the beautiful Roman goddess defined by sex, beauty, and fertility. Venus's Greek equivalent, Aphrodite, is defined by the same. Stories and teachings from Ancient Sumer reflect the same connection to the concept of a female deity.

From Egypt to the Mayans (and beyond), the significance of the luminous object known today as Venus can never be overstated. However, some cultures were unaware that the stars that appeared in the morning and the evening were one and the same. They connected stories to two beings rather than one.

An interesting fact is that on Venus a day is longer than a year. As odd as this sounds, one must think of it in celestial terms to understand it. A year simply means completing one trip around the sun (just as the Earth revolves around the sun in 365 days, 59 minutes, and 19 seconds). A day simply means a full rotation (or spin) around its own axis (just as the Earth is mere milliseconds from 24 hours). Venus takes just under 225 Earth days to complete a rotation around the sun, but 243 Earth days to rotate around its own axis.

Directions and More

Although Anishinaabe stories and teachings extend beyond the four commonly used directions (and are often acknowledged during times of offerings), these four directions can be very helpful in describing parts of the sky where you are observing.

The following directional names originated long ago from stories and teachings that described four sons (Waaban, Zhaawan, Ningaabii'an, and Giiwedin) born to Anishinaabe and a very special daughter of a firekeeper.

To the East – Waabanong
To the South – Zhaawanong
To the West – Ningaabii'anong
To the North – Giiwedinong

Exploring the inner workings of Anishinaabemowin can give further insights into the depth of the names of these directions (and their respective stories and teachings).